Africa's Water and Sanitation Infrastructure

Africa's Water and Sanitation Infrastructure

Access, Affordability, and Alternatives

Sudeshna Ghosh Banerjee and Elvira Morella

Vivien Foster and Cecilia Briceño-Garmendia,
Series Editors

THE WORLD BANK
Washington, D.C.

1818 H Street NW
Washington DC 20433
Telephone: 202-473-1000
Internet: www.worldbank.org

1 2 3 4 14 13 12 11

ISBN: 978-0-8213-8457-2
eISBN: 978-0-8213-8618-7
DOI: 10.1596/978-0-8213-8457-2

Library of Congress Cataloging-in-Publication Data
Africa's water and sanitation infrastructure: access, affordability, and alternatives / editors, Sudeshna Ghosh Banerjee, Elvira Morella.
p. cm.
Includes bibliographical references.
ISBN 978-0-8213-8457-2 — ISBN 978-0-8213-8618-7 (electronic)
1. Water utilities—Africa. 2. Water-supply—Economic aspects—Africa. 3. Sanitation—Economic aspects—Africa. 4. Sewage disposal—Economic aspects—Africa. I. Banerjee, Sudeshna Ghosh, 1973- II. Morella, Elvira, 1976- III. World Bank.
HD4465.A35.A47 2011
363.6'1096—dc22

2010047886

Cover photograph: Arne Hoel / The World Bank
Cover design: Naylor Design

Contents

Boxes

Figures

Tables

About the AICD

This study is a product of the Africa Infrastructure Country Diagnostic (AICD), a project designed to expand the world's knowledge of physical infrastructure in Africa. The AICD provides a baseline against which future improvements in infrastructure services can be measured, making it possible to monitor the results achieved from donor support. It also offers a more solid empirical foundation for prioritizing investments and designing policy reforms in the infrastructure sectors in Africa.

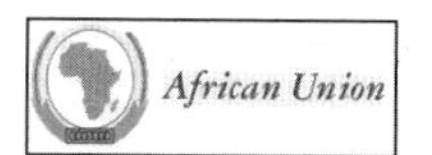

The AICD was based on an unprecedented effort to collect detailed economic and technical data on the infrastructure sectors in Africa. The project produced a series of original reports on public expenditure, spending needs, and sector performance in each of the main infrastructure sectors, including energy, information and communication technologies, irrigation, transport, and water and sanitation. The most significant findings were synthesized in a flagship report titled *Africa's Infrastructure: A Time for Transformation*. All the underlying data and models are available to the public through a Web portal (http://www.infrastructureafrica.org), allowing users to download customized data reports and perform various simulation exercises.

The AICD was commissioned by the Infrastructure Consortium for Africa following the 2005 G-8 Summit at Gleneagles, which flagged the importance of scaling up donor finance to infrastructure in support of Africa's development.

The first phase of the AICD focused on 24 countries that together account for 85 percent of the gross domestic product, population, and infrastructure aid flows of Sub-Saharan Africa. The countries were Benin, Burkina Faso, Cape Verde, Cameroon, Chad, Democratic Republic of Congo, Côte d'Ivoire,

Ethiopia, Ghana, Kenya, Lesotho, Madagascar, Malawi, Mozambique, Namibia, Niger, Nigeria, Rwanda, Senegal, South Africa, Sudan, Tanzania, Uganda, and Zambia. Under a second phase of the project, coverage was expanded to include the remaining countries on the African continent.

Consistent with the genesis of the project, the main focus was on the 48 countries south of the Sahara that face the most severe infrastructure challenges. Some components of the study also covered North African countries to provide a broader point of reference. Unless otherwise stated, therefore, the term "Africa" is used throughout this report as a shorthand for "Sub-Saharan Africa."

The AICD was implemented by the World Bank on behalf of a steering committee that represents the African Union, the New Partnership for Africa's Development (NEPAD), Africa's regional economic communities, the African Development Bank, and major infrastructure donors. Financing for the AICD was provided by a multidonor trust fund to which the main contributors were the Department for International Development (United Kingdom), the Public Private Infrastructure Advisory Facility, Agence Française de Développement, the European Commission, and Germany's Kreditanstalt für Wiederaufbau (KfW). The Sub-Saharan Africa Transport Policy Program and the Water and Sanitation Program provided technical support on data collection and analysis pertaining to their respective sectors. A group of distinguished peer reviewers from policy-making and academic circles in Africa and beyond reviewed all of the major outputs of the study to ensure the technical quality of the work.

Following the completion of the AICD project, long-term responsibility for ongoing collection and analysis of African infrastructure statistics was transferred to the African Development Bank under the Africa Infrastructure Knowledge Program (AIKP). A second wave of data collection of the infrastructure indicators analyzed in this volume was initiated in 2011.

Series Foreword

The Africa Infrastructure Country Diagnostic (AICD) has produced continent-wide analysis of many aspects of Africa's infrastructure challenge. The main findings were synthesized in a flagship report titled *Africa's Infrastructure: A Time for Transformation*, published in November 2009. Meant for policy makers, that report necessarily focused on the high-level conclusions. It attracted widespread media coverage feeding directly into discussions at the 2009 African Union Commission Heads of State Summit on Infrastructure.

Although the flagship report served a valuable role in highlighting the main findings of the project, it could not do full justice to the richness of the data collected and technical analysis undertaken. There was clearly a need to make this more detailed material available to a wider audience of infrastructure practitioners. Hence the idea of producing four technical monographs, such as this one, to provide detailed results on each of the major infrastructure sectors—information and communication technologies (ICT), power, transport, and water—as companions to the flagship report.

These technical volumes are intended as reference books on each of the infrastructure sectors. They cover all aspects of the AICD project relevant to each sector, including sector performance, gaps in financing and efficiency, and estimates of the need for additional spending on investment, operations, and maintenance. Each volume also comes with a detailed data appendix—providing easy access to all the relevant infrastructure indicators at the country level—which is a resource in and of itself.

In addition to these sector volumes, the AICD has produced a series of country reports that weave together all the findings relevant to one particular country to provide an integral picture of the infrastructure situation at the national level. Yet another set of reports provides an overall picture of the state of regional integration of infrastructure networks for each of the major regional economic communities of Sub-Saharan Africa. All of these papers are available through the project web portal, http://www.infrastructureafrica.org, or through the World Bank's Policy Research Working Paper series.

With the completion of this full range of analytical products, we hope to place the findings of the AICD effort at the fingertips of all interested policy makers, development partners, and infrastructure practitioners.

Vivien Foster and Cecilia Briceño-Garmendia

Acknowledgments

This book was coauthored by Sudeshna Ghosh Banerjee and Elvira Morella with support from Carolina Dominguez, under the overall guidance of series editors Vivien Foster and Cecilia Briceño-Garmendia. All are with the World Bank.

The book draws upon a number of background papers that were prepared by World Bank staff and consultants, under the auspices of the Africa Infrastructure Country Diagnostic (AICD). Key contributors to the book on a chapter-by-chapter basis were as follows.

Chapter 1

Contributors

Sudeshna Ghosh Banerjee, Cecilia Briceño-Garmendia, Tarik Chfadi, Amadou Diallo, Carolina Dominguez, Vivien Foster, Sarah Keener, Manuel Luengo, Elvira Morella, Taras Pushak, Heather Skilling, Clarence Tsimpo, Helal Uddin, Quentin Wodon.

Key Source Documents

Banerjee, S., H. Skilling, V. Foster, C. Briceño-Garmendia, E. Morella, and T. Chfadi. 2008. "Ebbing Water, Surging Deficits: Urban Water Supply in Sub-Saharan Africa." AICD Background Paper 12, World Bank, Washington, DC.

Banerjee, S., Q. Wodon, A. Diallo, T. Pushak, H. Uddin, C. Tsimpo, and V. Foster. 2008. "Access, Affordability and Alternatives: Modern Infrastructure Services in Sub-Saharan Africa." AICD Background Paper 2, World Bank, Washington, DC.

Briceño-Garmendia, C., K. Smits, and V. Foster. 2008. "Financing Public Infrastructure in Sub-Saharan Africa: Patterns and Emerging Issues." AICD Background Paper 15, World Bank, Washington, DC.

Keener, S., M. Luengo, and S. G. Banerjee. 2009. "Provision of Water to the Poor in Africa: Experience with Water Standposts and the Informal Water Sector." AICD Working Paper 13, World Bank, Washington, DC.

Morella, E., V. Foster, and S. Banerjee. 2008. "Climbing the Ladder: The State of Sanitation in Sub-Saharan Africa." AICD Background Paper 13, World Bank, Washington, DC.

Chapter 2

Contributors

Sudeshna Ghosh Banerjee, Cecilia Briceño-Garmendia, Tarik Chfadi, Amadou Diallo, Vivien Foster, Sarah Keener, Manuel Luengo, Elvira Morella, Taras Pushak, Heather Skilling, Clarence Tsimpo, Helal Uddin, Quentin Wodon.

Key Source Documents

Banerjee, S., H. Skilling, V. Foster, C. Briceño-Garmendia, E. Morella, and T. Chfadi. 2008. "State of the Sector Review: Rural Water Supply." AICD Working Paper, World Bank, Washington, DC.

Banerjee, S., Q. Wodon, A. Diallo, T. Pushak, H. Uddin, C. Tsimpo, and V. Foster. 2008. "Access, Affordability and Alternatives: Modern Infrastructure Services in Sub-Saharan Africa." AICD Background Paper 2, World Bank, Washington, DC.

Keener, S., M. Luengo, and S. G. Banerjee. 2009. "Provision of Water to the Poor in Africa: Experience with Water Standposts and the Informal Water Sector." AICD Working Paper 13, World Bank, Washington, DC.

Chapter 3

Contributors

Elvira Morella, Sudeshna Ghosh Banerjee, Amadou Diallo, Vivien Foster, Taras Pushak, Clarence Tsimpo, Helal Uddin, Quentin Wodon.

Key Source Documents

Banerjee, S., Q. Wodon, A. Diallo, T. Pushak, H. Uddin, C. Tsimpo, and V. Foster. 2008. "Access, Affordability and Alternatives: Modern Infrastructure Services in Sub-Saharan Africa." AICD Background Paper 2, World Bank, Washington, DC.

Morella, E., V. Foster, and S. Banerjee. 2008. "Climbing the Ladder: The State of Sanitation in Sub-Saharan Africa." AICD Background Paper 13, World Bank, Washington, DC.

Chapter 4

Contributors

Maria Vagliasindi, Cecilia Briceño-Garmendia, Tarik Chfadi, Vivien Foster, Sarah Keener, Manuel Luengo, John Nellis, Heather Skilling, Sudeshna Ghosh Banerjee.

Key Source Documents

Vagliasindi, M., and J. Nellis. 2009. "Evaluating Africa's Experience with Institutional Reforms for the Infrastructure Sectors." AICD Working Paper 23, World Bank, Washington, DC.

Banerjee, S., H. Skilling, V. Foster, C. Briceño Garmendia, E. Morella, and T. Chfadi. 2008. "Ebbing Water, Surging Deficits: Urban Water Supply in Sub-Saharan Africa." AICD Background Paper 12, World Bank, Washington, DC.

Keener, S., M. Luengo, and S. G. Banerjee. 2009. "Provision of Water to the Poor in Africa: Experience with Water Standposts and the Informal Water Sector." AICD Working Paper 13, World Bank, Washington, DC.

Chapter 5

Contributors

Cecilia Briceño-Garmendia, Sudeshna Ghosh Banerjee, Tarik Chfadi, Vivien Foster, John Nellis, Heather Skilling, Karlis Smits, Maria Vagliasindi, Quentin Wodon, Yvonne Ying.

Key Source Documents

Banerjee, S., V. Foster, Y. Ying, H. Skilling, and Q. Wodon. 2008. "Cost Recovery, Equity and Efficiency in Water Tariffs: Evidence from African Utilities." AICD Working Paper 7, World Bank, Washington, DC.

Banerjee, S., H. Skilling, V. Foster, C. Briceño-Garmendia, E. Morella, and T. Chfadi. 2008. "Ebbing Water, Surging Deficits: Urban Water Supply in Sub-Saharan Africa." AICD Background Paper 12, World Bank, Washington, DC.

Briceño-Garmendia, C., K. Smits, and V. Foster. 2008. "Financing Public Infrastructure in Sub-Saharan Africa: Patterns and Emerging Issues." AICD Background Paper 15, World Bank, Washington, DC.

Vagliasindi, M., and J. Nellis 2009. "Evaluating Africa's Experience with Institutional Reforms for the Infrastructure Sectors." AICD Working Paper 23, World Bank, Washington, DC.

Chapter 6

Contributors

Vivien Foster, Sudeshna Ghosh Banerjee, Tarik Chfadi, Amadou Diallo, Sarah Keener, Manuel Luengo, Taras Pushak, Heather Skilling, Clarence Tsimpo, Helal Uddin, Quentin Wodon, Yvonne Ying.

Key Source Documents

Banerjee, S., V. Foster, Y. Ying, H. Skilling, and Q. Wodon. 2008. "Cost Recovery, Equity and Efficiency in Water Tariffs: Evidence From African Utilities." AICD Working Paper 7, World Bank, Washington DC.

Banerjee, S., Q. Wodon, A. Diallo, T. Pushak, H. Uddin, C. Tsimpo, and V. Foster. 2008. "Access, Affordability and Alternatives: Modern Infrastructure Services in Sub-Saharan Africa." AICD Background Paper 2, World Bank, Washington, DC.

Keener, S., M. Luengo, and S. G. Banerjee. 2009. "Provision of Water to the Poor in Africa: Experience with Water Standposts and the Informal Water Sector." AICD Working Paper 13. World Bank, Washington, DC.

Morella, E., V. Foster, and S. Banerjee. 2008. "Climbing the Ladder: The State of Sanitation in Sub-Saharan Africa." AICD Background Paper 13, World Bank, Washington, DC.

Chapter 7

Contributors

Elvira Morella, Africon.

Key Source Documents

Africon. 2008. "Unit Costs of Infrastructure Projects in Sub-Saharan Africa." AICD Background Paper 11, World Bank, Washington, DC.

Chapter 8

Contributors

Cecilia Briceño-Garmendia, Carolina Dominguez, Vivien Foster, Nataliya Pushak, Karlis Smits.

Key Source Documents

Briceño-Garmendia, C., K. Smits, and V. Foster. 2008. "Financing Public Infrastructure in Sub-Saharan Africa: Patterns and Emerging Issues." AICD Background Paper 15, World Bank, Washington, DC.

Foster, Vivien, William Butterfield, Chuan Chen, and Nataliya Pushak. 2008. "Building Bridges: China's Growing Role as Infrastructure

Financier for Sub-Saharan Africa." Trends and Policy Options 5, Public-Private Infrastructure Advisory Facility, World Bank, Washington, DC.

Chapter 9

This chapter is a synthesis of our findings.

None of this research would have been possible without the generous collaboration of government officials in the key sector institutions of each country, as well as the arduous work of local consultants who assembled this information in a standardized format. In addition, thanks are due to the staff of the Water and Sanitation Program (WSP), who have collaborated in the data collection process and rallied to support this endeavor in their respective countries by facilitating meetings, sharing documentation, and completing data collection where needed. Dennis Mwanza, former urban water supply thematic leader at WSP-Africa, has been an active partner in supporting this process from its inception. Piers Cross, Vivian Castro, Valentina Zuin, and Jean Doyen, also of WSP-Africa, have been invaluable partners in this effort, ensuring that messages are aligned with real-world knowledge of the sector and represent the on-the-ground realities in Africa.

Finally, the project team is grateful to the team of local consultants, sourced from PricewaterhouseCoopers-Africa (PwC) and independent local consultants, who carried out this exercise in the field amid all the problems associated with gathering data at this scale. A list of these partners appears below. Afua Sarkodie ably led the PwC team and provided oversight and support to the local consultants.

Country	*Collaborating institutions (Water and Sanitation Program, the World Bank)*	*Local consultants or other partners*
Benin	Sylvain Migan	
Burkina Faso	Seydou Traore, Christophe Prevost	
Cameroon	Astrid Manroth	
Cape Verde		Sandro de Brito
Chad	Yao Badjo	Kenneth Simo
Congo, Dem. Rep.	Georges Kazad	Henri Kabeya
Côte d'Ivoire	Emmanuel Diarra	Eric Boa

(continued next page)

Country	*Collaborating institutions (Water and Sanitation Program, the World Bank)*	*Local consultants or other partners*
Ethiopia	Belete Muluneh	Yemarshet Yemaneh Mengistu
Ghana	Ventura Bengoechea	Afua Sarkodie
Kenya	Dennis Mwanza, Japheth Mbuvi, Vivian Castro	Ayub Osman, Peter Njui
Lesotho	Jane Walker	Peter Ramsden
Madagascar	Christophe Prevost	Gerald Razafinjato
Malawi	Midori Makino, Bob Roche	Caroline Moyo
Mozambique	Jane Walker, Luiz Tavares, Valentina Zuin	Carla Barros Costa
Namibia		Birgit de Lange, Peter Ramsden
Niger	Ibrah Sanoussi, Matar Fall	
Nigeria	Joe Gadek, Hassan Kida	Mohammed Iliyas
Rwanda	Bruno Mwanafunzi, Christophe Prevost	Charles Uramutse
Senegal	Pierre Boulanger	Ndongo Sene
South Africa		Peter Ramsden
Sudan	Solomon Alemu	A. R. Mukhtar
Tanzania	Francis Ato Brown, Nat Paynter	Kenneth Simo
Uganda	Samuel Mutono	
Zambia	Barbara Senkwe	Caroline Moyo

The work benefited from widespread peer review by colleagues within the World Bank, notably Robert Roche, Alexander McPhail, Caroline van den Berg, Christophe Prevost, Ventura Bengoechea, Dennis Mwanza, and Meike van Ginneken. The external peer reviewer for this volume, Sophie Tremolet, provided constructive and thoughtful comments. The comprehensive editorial effort of Steven Kennedy is much appreciated.

Abbreviations

ADAMA	Nazareth Water Company, Ethiopia
ADeM	Águas de Moçambique
AICD	Africa Infrastructure Country Diagnostic
AREQUAPCI	association of water resellers
AWSA	Addis Ababa Water Services Authority, Ethiopia
BOCC	basket of construction components
BWB	Blantyre Water Board, Malawi
CAPEX	capital expenditure
CFA	Communauté Financière Africaine Franc
CBO	community-based organization
CEMAC	Central African Economic and Monetary Community
COMESA	Common Market for Eastern and Southern Africa
CRWB	Central Region Water Board, Malawi
DAWASCO	Dar es Salaam Water and Sewerage Company, Tanzania
DBT	direct block tariff
DHS	demographic and health survey
DUWS	Dodoma Urban Water and Sewerage Authority, Tanzania
EAC	East African Community
ECOWAS	Economic Community of West African States
FCT	Federal Capital Territory, Nigeria

GDP	gross domestic product
GNI	gross national income
GRUMP	Global Rural-Urban Mapping Project
GWC	Ghana Water Company
HCI	high conflict index
IBNET	International Benchmarking Network for Water and Sanitation Utilities
IBT	increasing block tariff
ICP	International Comparison Program
IDA	International Development Association
IDAMC	Internally Delegated Area Management Contract
IFRS	International Financial Reporting Standards
JIRAMA	Jiro sy Rano Malagasy, Madagascar
JMP	Joint Monitoring Programme
KIWASCO	Kisumu Water and Sewerage Company, Kenya
LCI	low conflict index
LWB	Lilongwe Water Board, Malawi
LWSC	Lusaka Water and Sewerage Company, Zambia
MCI	medium conflict index
MDG	Millennium Development Goal
MICS	multiple-indicator cluster survey
MSNE	Mauritania Société Nationale d'Eau et d'Electricité
MWSA	Mwanza Water and Sewerage Authority, Tanzania
MWSC	Mombasa Water and Sewerage Company, Kenya
NGO	nongovernmental organization
NRW	nonrevenue water
NWASCO	Nairobi Water and Sanitation Company, Kenya
NWC	National Water Company
NWSC	National Water and Sewerage Company, Uganda
NWSC	Nkana Water and Sewerage Company, Zambia
O&M	operations and maintenance
ODA	official development assistance
OECD	Organisation for Economic Co-operation and Development
ONAS	Office National de l'Assainissement du Sénégal
ONEA	Office Nationale des Eaux et d'Assainissement, Burkina Faso
OPEX	operating expenditure
PPI	private participation in infrastructure
PPIAF	Public-Private Infrastructure Advisory Facility

PPP	purchasing power parity
PSP	private sector participation
PwC	PricewaterhouseCoopers-Africa
REGIDESO	Régie de Production et de Distribution d'Eau
RUWATSSA	State Rural Water Supply and Sanitation Agency, Nigeria
SADC	Southern African Development Community
SDE	Sénégalaise des Eaux
SEEG	Société d'Electricité et d'Eaux du Gabon
SEEN	Société de Exploitation des Eaux du Niger
SNEC	Société National des Eaux du Cameroon
SODECI	Société de Distribution d'Eau de Côte d'Ivoire
SOE	state-owned enterprise
SONEB	Société Nationale des Eaux du Benin
SPEN	Société de Patrimoine des Eaux du Niger
STEE	Société Tchadienne d'Eau et d'Electricité, Chad
SWSC	Southern Water and Sewerage Company, Zambia
TdE	Togolaise des Eaux
UNICEF	United Nations Children's Fund
VIP	ventilated improved pit
WASA	Water and Sanitation Authority, Lesotho
WB	Water Board
WHO	World Health Organization
WSP	Water and Sanitation Program
WSP-SA	Water and Sanitation Program–South Asia
WSS	water supply and sanitation
WUC	Water Utilities Corporation, Botswana

CHAPTER 1

The Elusiveness of the Millennium Development Goals for Water and Sanitation

The welfare implications of safe water cannot be overstated. Infectious diarrhea and other serious waterborne illnesses are leading causes of infant mortality and malnutrition. Their impact extends beyond health to the economic realm in the form of lost work days and school absenteeism. It is estimated that meeting the Millennium Development Goal (MDG) for access to safe water[1] would produce an economic benefit of US$3.1 billion (in 2000 dollars) in Africa, a gain realized by a combination of time savings and health benefits. The cost-benefit ratio is about 11, which suggests that the benefits derived from access to safe water are far greater than the costs of providing it (Hutton and Haller 2004).

Similarly, sanitation makes a key contribution to public health, particularly in densely populated areas. Adequate sanitation is defined as any private or shared, but not public, facility that guarantees that waste is hygienically separated from human contact (JMP 2000). Adequate sanitation reduces the risk of a broad range of diseases—including respiratory ailments, malaria, and diarrhea—and reduces the prevalence of malnutrition. Access to this standard of sanitation produces direct health gains by preventing disease and delivering economic and social benefits. It is estimated that a reduction in diarrheal illness would produce a gain of 99 million days of school and 456 million days of work for the working population ages 15–59 in Africa. The workdays alone represent economic benefits equal to as much as US$116 million (Hutton and Haller 2004).

The international adoption of the MDGs in 2000 created a new framework for focusing poverty reduction efforts on the indicators that are most meaningful for economic development. The MDGs have called attention to deficiencies in the quantity and quality of water supply and sanitation (WSS). MDG 7 calls for ensuring environmental sustainability and—relevant to this book—reducing by half the number of people without sustainable access to safe drinking water and improved sanitation. Although the world overall is on track to meet the MDG drinking water target, Africa lags. The gap is most acute in Sub-Saharan Africa, where only 58 percent of the population enjoys access to safe drinking water, and the gap is widening, as the increasingly urban population places a greater strain on existing service providers (table 1.1). Of the 828 million people in the world whose water sources remain unimproved, 37 percent live in Sub-Saharan Africa. According to projections, 300 million people—almost 38 percent of Sub-Saharan Africa's population, or half the number of people who presently have access to improved water—will need to be covered to meet the MDG target (JMP 2008).

The world is not on track to meet the MDG sanitation target. More than 2.5 billion people remain without improved sanitation worldwide; of that total, 22 percent, or more than half a billion people, live in Africa. A reported 221 million people in Africa still defecate in the open, the second-largest total for any region after South Asia. Access to improved sanitation

Table 1.1 Regional Progress toward the MDG Drinking Water Target

	Drinking water coverage (%)				
Region	*1996*	*2006*	*Coverage needed to be on track in 2006 (%)*	*MDG target coverage (%)*	*Progress*
Sub-Saharan Africa	49	58	65	75	Off track
North Africa	88	92	92	94	On track
Latin America and the Caribbean	84	92	89	92	On track
East Asia	68	88	78	84	On track
South Asia	74	87	82	87	On track
Southeast Asia	73	86	82	87	On track
West Asia	86	90	90	93	On track

Source: JMP 2008.

Table 1.2 Regional Progress toward the MDG Sanitation Target

Region	*Sanitation coverage (%)*		*Coverage needed to be on track in 2006 (%)*	*MDG target coverage (%)*	*Progress*
	1990	*2006*			
West Asia	79	84	86	90	On track
Latin America and the Caribbean	68	79	78	84	On track
North Africa	62	76	74	81	On track
Southeast Asia	50	67	64	75	On track
East Asia	48	65	65	74	On track
South Asia	21	33	46	61	Off track
Sub-Saharan Africa	26	31	50	63	Off track
World	54	62	69	77	Off track

Source: JMP 2008.

has increased only modestly in Sub-Saharan Africa, from 26 percent of the total population in 1990 to 31 percent in 2006. To be on track with the MDG's sanitation benchmark, improved sanitation coverage should have been at 50 percent of the population in 2006. To meet the MDG sanitation target, the current number of people with improved sanitation in Africa needs to more than double, from 242 million in 2006 to 615 million in 2015. Unless the current trend changes, Sub-Saharan Africa will definitely not meet the sanitation target (table 1.2).

A Timely Synthesis

With only five years remaining until the MDG deadline in 2015, it is essential to take stock of the status of the WSS sectors, analyze their achievements and shortcomings in Sub-Saharan Africa, and identify the sector characteristics that either advance or inhibit the population's ability to access service. Governments have adopted WSS reforms and attracted investments to build dynamism in the sectors and to enhance performance outcomes. These initiatives have been critical to developing implementation capacity and to establishing innovative forms of service delivery.

Building on background work carried out under the auspices of the Africa Infrastructure Country Diagnostic (AICD) and presented by Foster and Briceño-Garmendia (2009), this volume integrates a wealth of primary and secondary information to present a quantitative snapshot of the state of the WSS sectors in Africa, including the current status of access and coverage trends. It explains institutional and governance structures and utility performance and articulates the volume and quality of financing available over time for WSS. The volume also evaluates the challenges to the WSS sectors and explores the factors that might explain the expansion of coverage. Finally, it endeavors to estimate spending needs for WSS, with a target of meeting the MDG goal, and compares those needs with the existing financing envelopes, disaggregated into shares that can be recouped through efficiency improvements and gaps that would remain even if all feasible efficiencies were achieved. The directions for the future draw on lessons learned from experiences around the continent and present the menu of choices available to African countries.

Data Sources and Methodologies

Monitoring the progress of infrastructure sectors such as water supply has been a significant by-product of the MDGs, and serious attention and funding have been devoted in recent years to developing monitoring and evaluation systems in countries around the world. The Joint Monitoring Programme (JMP) on WSS is an institutional endeavor by the World Health Organization and the United Nations Children's Fund to systematically track progress toward the WSS MDG. The JMP's monitoring introduced the concept of improved and unimproved WSS and categorized WSS sources according to the typology shown in table 1.3.

The JMP and AICD Methodologies

AICD used a body of household surveys similar to that of the JMP—demographic and health surveys (DHSs), multiple-indicator cluster surveys (MICSs), and income/expenditure surveys—but the JMP has adopted special rules for use when the exact disaggregation across WSS modalities is not available in the surveys. Those rules apply most often to the largest sources of WSS, namely, wells or boreholes and traditional pit latrines. The JMP statistics apportion 50 percent of wells or boreholes to the protected or "improved" category and the remainder to the unprotected or "unimproved" category. Similarly, covered pit latrines are placed in the "improved" category, and the unprotected in the "unimproved" category. In the AICD analysis, the information available in the survey has been taken at face value without any adjustment. Therefore, only the household connections to piped water and piped water delivered through public standposts constitute the "improved water" category, and flush toilets and ventilated improved pit (VIP) latrines are included in the

Table 1.3 Definition of Coverage of Improved Water

	JMP category	*AICD category*
Primary source of water supply		
Piped water into dwelling or yard	Improved	Improved
Public tap or communal standpipe, standposts, or kiosks	Improved	Improved
Wells or boreholes, hand pumps, or rainwater	Improved/unimproved	Unimproved
Surface water (for example, lake, river, pond, dam, or spring)	Unimproved	Unimproved
Vendors or tanker trucks	Unimproved	Unimproved
Other (for example, bottled water)	Unimproved	Unimproved
Primary source of sanitation		
Flush toilet to network or septic tank	Improved	Improved
VIP latrine, SanPlat, or basic pits with slab	Improved	Improved
Traditional pit latrine	Improved/unimproved	Unimproved
Bucket or other container	Unimproved	Unimproved
Other	Unimproved	Unimproved
No facility (nature or bush)	Unimproved	Unimproved

Source: Banerjee, Wodon, and others 2008.
Note: VIP = ventilated improved pit.

"improved sanitation" category. Further, the DHSs describe access to sanitation without discriminating between on-site sanitation and use of sewerage facilities, so that both are included in the flush toilet category. Most of these flush toilets, however, use septic tanks rather than sewer connections. For this reason, this study assumes that the DHS information relating to flush toilets refers to septic tanks.

Owing to these methodological differences, the JMP and AICD figures differ on improved water and improved sanitation. Not surprisingly, the differences are more pronounced in rural areas, where wells/boreholes and traditional pit latrines are the most prevalent forms of WSS sources (figure 1.1). In this volume, we focus above all on what lies within each of the improved and unimproved categories, rather than on the aggregates. Further, JMP uses methodologies that usually differ from methodologies used by each country to evaluate coverage. In most cases, national statistics would show higher coverage figures than does JMP.

Sanitation can be provided on numerous distinct levels that can be graphically represented as rungs on a ladder. Starting from open defecation, the successive increments are traditional latrines (various kinds of pits), improved latrines (including SanPlat, VIP latrines, and basic pits

Figure 1.1 JMP and AICD Estimates of the Prevalence of "Improved" Water Supply and Sanitation

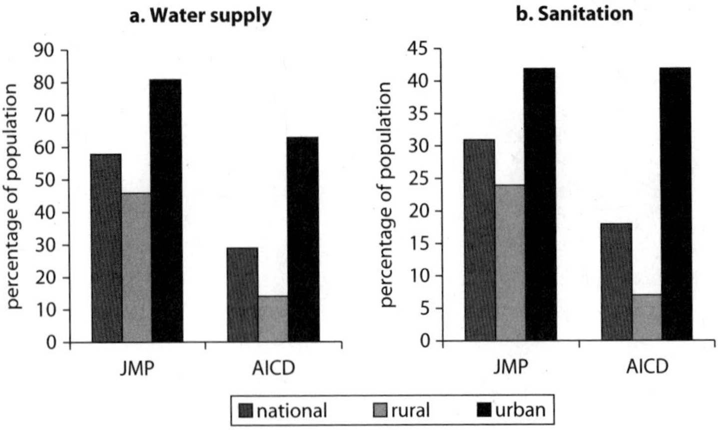

a. Water supply **b. Sanitation**

national rural urban

Sources: Banerjee, Wodon, and others 2008; JMP 2008.

with slabs), and flush toilets (connected to either a septic tank or a waterborne sewage network). The higher rungs of the ladder carry higher unit costs and lower levels of perceived health risk (figure 1.2).

This concept carries over to water, but not as clearly, because the sources cannot be ranked on the basis of quality or cost. It is evident, however, that surface water represents the bottom rung, and household connections to piped water and piped water delivered through public standposts are at the upper end of the ladder. What comes out very clearly in the literature is that the distance to the water source makes a substantial difference to health outcomes and time savings.

Data Sources

The analysis presented in this book is based on three primary databases that underlie three AICD background papers—Banerjee, Skilling, and others (2008), Banerjee, Wodon, and others (2008), and Morella, Foster, and Banerjee (2008). These background papers are referred to throughout this volume.

Household surveys: AICD DHS/MICS Database and Expenditure Database. The results from household surveys are used extensively in chapters 2 and 3. The first, the AICD DHS/MICS database, was used to analyze the current status and access trends presented in this volume; it is a composite of 63 DHS and MICS data sets. Thirty countries in Africa

Figure 1.2 The Sanitation Ladder

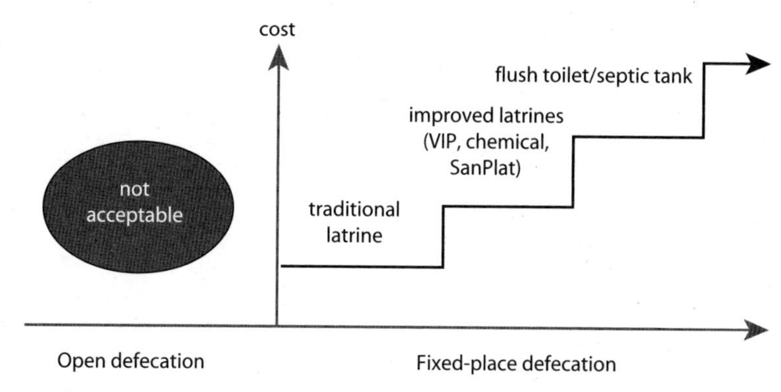

Source: Authors.
Note: VIP = ventilated improved pit.

have had at least one DHS conducted since 1990, and 24 have at least two DHS data points between 1990 and 2005, which enables trend analysis. Second, the AICD expenditure survey database includes the most recent household-level expenditure surveys for 30 African countries during the period from 1997 to 2005. This database incorporates surveys modeled after the Living Standards Measurement Surveys. These surveys provide a wealth of information on use of and payment for infrastructure services, as well as offering data on household assets and expenditure patterns. Known by different names in different countries, these surveys are carried out by country governments to reflect local nuances and priorities. Therefore, their infrastructure modules often are not harmonized or comparable, and standardization techniques have been employed to permit continentwide inferences (annexes 1.1 and 1.2).

AICD Water Supply and Sanitation Survey. This survey was carried out in two phases and administered to line ministries, sector institutions, and water utilities with a view to capturing institutional and performance variables. Seven modules of data were collected for each country, of which five are qualitative and two are quantitative. The focus of each module is reflected in table 1.4.

The data were collected in two phases (2007 and 2009) and from two distinct sources (AICD and the International Benchmarking Network for Water and Sanitation Utilities [IBNET]). AICD's data collection in 24 countries in 2007 resulted in a comprehensive data set covering 51 utilities. AICD's 2009 flagship report (Foster and Briceño-Garmendia 2009) was

Table 1.4 Modules of AICD WSS Survey

Module	*Description*	*Data collection unit*	*Data collection source and coverage*	*Topics in questionnaire*
Module 1: Institutional and regulatory	Qualitative	Country	AICD Phase I	Legal framework, sector organization, regulatory framework, regulatory process, tariff adjustment, private participation
Module 2: Rural water	Qualitative	Country	AICD Phase I	Sector organization, service characteristics
Module 3: Governance	Qualitative	Utility	AICD Phase I	Ownership, board structure, performance contract, performance monitoring and disclosure, finance, labor
Module 4: Sanitation	Qualitative	Country	AICD Phase I	Sector organization, service characteristics
Module 5: Small-scale independent providers	Qualitative	Largest city	AICD Phase I	Point sources, mobile sources
Module 6: Operational and financial	Quantitative	Utility	AICD Phase I, II, IBNET	Access, quality of service, operational performance, financial performance
Module 7: Tariff schedules	Quantitative	Utility	AICD Phase I	Currently effective tariff schedule

Source: Authors.
Note: IBNET = International Benchmarking Network for Water and Sanitation Utilities.

based on this data-collection effort. In 2009, AICD carried out a second round of data collection in three more countries and covering three additional utilities, but this information was restricted to operational and financial performance only (module 6). The AICD data set was integrated with that of IBNET, which collected operational and financial performance data (module 6) for 32 more utilities. The upper bound of the data set covers 32 countries and 86 utilities; the lower bound covers 24 countries and 51 utilities.

Different modules underpin the individual chapters in this volume. For instance, chapter 2 draws on modules 2 and 5 to elaborate on the current state of the formal, informal, and rural water markets. Chapter 3 employs the questions in module 4 to present the sanitation snapshot. Chapter 4 draws on modules 1 and 3, which contain questions detailing the institutional environment of the WSS sectors. Quantitative data were captured to develop an understanding of the financial, technical, and operational performance of the selected utilities (module 6). Utilities were asked to provide data for the 10-year period from 1995 to 2005, but because older data were rarely available, the emphasis shifted to collecting data from the five-year period from 2000 to 2005. Chapter 5 is based on the operational and financial time-series data on utility performance contained in module 6. The information presented in tariff schedules (module 7) is used in chapters 5 and 6.

AICD fiscal database. The country-level analysis of the volumes, patterns, and composition of financial resources for WSS draws on the AICD fiscal database used extensively in chapter 8. That database, which captures information on public spending in the infrastructure sectors of 25 countries, is a unique attempt to document in a standardized manner the levels and patterns of public spending for infrastructure, including WSS. If one uses the database, it is possible to comparisons across sectors and ensure consistency over time. Financing flows within public spending are defined as including tax revenue or user charges channeled through both on-budget (central and local governments) and off-budget mechanisms (state-owned enterprises and special funds).

Country Categories

The performance of the utilities is evaluated across various functional and financial dimensions and presented in selected country groupings consistent with the method used in Foster and Briceño-Garmendia (2009), described in annex 1.3. The country groupings are based on (1) income and fragility (middle-income, low-income fragile, low-income,

Table 1.5 Utilities Analyzed in This Report, by Categories

	IBNET	*AICD*	*Total*
Sub-Saharan Africa	32	54	86
Income group			
Low-income, fragile	1	2	3
Low-income, nonfragile	20	26	46
Middle-income	1	11	12
Resource-rich	10	15	25
Regional economic community			
CEMAC	1	3	4
COMESA	10	19	29
EAC	17	8	
ECOWAS	3	14	
SADC	27	26	
Water availability			
High water scarcity	2	30	32
Low water scarcity	30	24	54

Source: Authors.
Note: CEMAC = Central African Economic and Monetary Community; COMESA = Common Market for Eastern and Southern Africa; EAC = East African Community; ECOWAS = Economic Community of West African States; SADC = Southern African Development Community.

and resource-rich), (2) water scarcity (high, low), and (3) regional economic community (EAC, ECOWAS, CEMAC, COMESA, and SADC) (table 1.5). The utilities are further distinguished by size (small, large).

Key Finding 1: Wide Differences in Patterns of Access to Water

In rural areas, reliance on surface water remains prevalent, and boreholes are the principal improved source of water. The share of the population relying on surface water fell sharply in the 1990s, from 50 percent to just more than 40 percent, where it has remained for the past five years (table 1.6). Boreholes are the main source of improved water, accounting for a further 40 percent of the population. Access to piped water and standposts is very low, barely increasing over the period 1990–2005. Indeed, in many countries, less than 1 percent of the rural population receives piped water. It is striking that in more urbanized countries, access to piped water and standposts in rural areas is substantially higher.

In urban areas, coverage of piped water fell markedly over the period 1990–2005 owing to rapid population growth. At close to 40 percent, however, it is still the single largest source of urban water. Coverage of

Table 1.6 Evolution of Water Supply Coverage in Africa, by Source
(percent)

	Piped supply		*Standposts*		*Well and boreholes*		*Surface water*	
Period	*Urban*	*Rural*	*Urban*	*Rural*	*Urban*	*Rural*	*Urban*	*Rural*
1990–95	50	4	29	9	20	41	6	50
1995–2000	43	4	25	9	21	41	5	41
2001–05	39	4	24	11	24	43	7	42

Source: Banerjee, Wodon, and others 2008.

standposts saw a similar decline, but that of boreholes rose, so that each represented about 24 percent of the urban population in 2005. Overall, about two-thirds of the urban populace depends on utility water. The lower coverage of standposts compared with piped water is particularly striking, given the relatively low cost of standposts and the pressure to expand services rapidly. Reliance on surface water, at 7 percent of the urban population, changed little between 1990 and 2005.

Utilities are the central actors responsible for water supply in urban areas. In the middle-income countries they are essentially the only players, reaching about 98 percent of the urban population, the vast majority through private piped-water connections. In low-income countries only 68 percent of urban residents benefit from utility water, fewer than half through private piped connections (table 1.7). For the rest, informal sharing of connections through resale between neighbors (15 percent of the urban population) is almost as prevalent as formal sharing through standposts (19 percent of the urban population).

Utilities report providing about 20 hours per day of service (table 1.8). They typically produce just more than 200 liters per customer served, though the amount for middle-income countries is about twice that for low-income countries. If the total water production of the utilities could be evenly distributed to the entire population residing in the utility service area, it would amount to 74 liters per capita a day, just about adequate to meet basic human needs.

Urban households that do not benefit from utility water rely on several alternatives. The rapid expansion of boreholes in urban areas has already been noted. Water vendors, another alternative, may sell water obtained from utilities, boreholes, or surface sources from either trucks and carts or, less frequently, through private distribution networks. Water vendors account for only 3 percent of the African urban market,

Table 1.7 Services Provided by Utilities in Their Service Areas
(percent)

	Access by private residential piped-water connection	*Access by standpost*	*Access by sharing neighbors' private connection*	*Access to utility water by some modality*
Sub-Saharan Africa	44.3	13.0	21.7	64.0
Low-income countries	42.2	23.2	22.5	68.6
Low-income, fragile countries	25.6	2.2	41.0	56.0
Resource-rich countries	30.3	15.8	7.4	48.8
Middle-income countries	88.0	9.7	0.3	97.8

Source: Banerjee, Skilling, and others 2008.

Table 1.8 Quality of Services Provided by Utilities in Their Service Areas

	Availability of water		*Quality of supply*	
	Water production per resident in the utility service area (liters per capita per day)	*Water production per customer served by utility in service area (liters per capita per day)*	*Samples passing chlorine test (%)*	*Continuity of water service (hours per day)*
Sub-Saharan Africa	116.4	162.9	87.9	19.6
Low-income countries	66.0	130.2	92.8	19.0
Low-income, fragile countries	35.7	76.5	75.3	18.2
Resource-rich countries	140.5	208.8	78.1	18.4
Middle-income countries	208.8	233.6	97.2	24.0

Source: Banerjee, Skilling, and others 2008.

rising to 7 percent in West Africa. In some countries, however, their contribution to urban water supply is much larger: Nigeria (10 percent), Chad (16 percent), Niger (21 percent), and Mauritania (32 percent). In 15 large cities in Africa, the cost of vendor water, particularly when transported directly to the household, can be 2–11 times more expensive than having a household connection (table 1.9). This high willingness to pay for vendor water is a potential revenue source that the utilities are typically unable to capture.

Wells and boreholes are by far the fastest-growing source of improved water in both urban and rural areas. Service expansion shows a similar overall pattern in both cases: The absolute number of people depending

Table 1.9 Average Price for Water Service in 15 Largest Cities, by Type of Provider

	House connection	*Small piped network*	*Standpost*	*Household reseller*	*Water tanker*	*Water vendor*
Average price (US$ per cubic meter)	0.49	1.04	1.93	1.63	4.67	4.00
Markup over house connection (%)	100	214	336	402	1,103	811

Source: Keener, Luengo, and Banerjee 2009.

on surface water continues to grow, a grim statistic in its own right (figure 1.3). Across the board, wells and boreholes are expanding coverage much more rapidly than all the utility-based alternatives put together. Within the purview of the utility, access to standposts seems to be growing faster than piped water. However, the combined growth rates of the various improved forms of water in urban areas (less than 1 percent a year) still fall short of population growth (more than 4 percent a year).

Access to improved water sources is highly inequitable across the income distribution (figure 1.4). Access to piped water and standposts is heavily concentrated among the more affluent segments of the population, typically in urban areas. The poorest 40 percent of the population, by contrast, depends on surface water and on wells and boreholes in almost equal measure. Only 10 percent of African households in the bottom 60 percent of population are covered by piped supply. For the middle-income countries, access to piped water and standposts among the poorest quintiles is substantially higher than in the low-income countries.

Key Finding 2: Equally Wide Differences in Patterns of Access to Sanitation

Traditional pit latrines are by far the most common facility in both urban and rural areas, but more than a third of the population—mostly in rural areas—still defecates in the open (table 1.10). Improved sanitation (septic tanks and improved latrines) reaches less than 20 percent of Africa's population, and less than 10 percent in rural areas. Coverage of improved latrines is no greater than that of septic tanks, despite the significant cost difference between them. Only 10 percent of the population uses a septic tank; coverage in rural areas is practically negligible. In urban areas, septic tanks are much more common than improved latrines, and less than 10 percent of the population practices open defecation.

Figure 1.3 Dependence on Surface Water in Urban and Rural Areas, 1990s versus Early 2000s

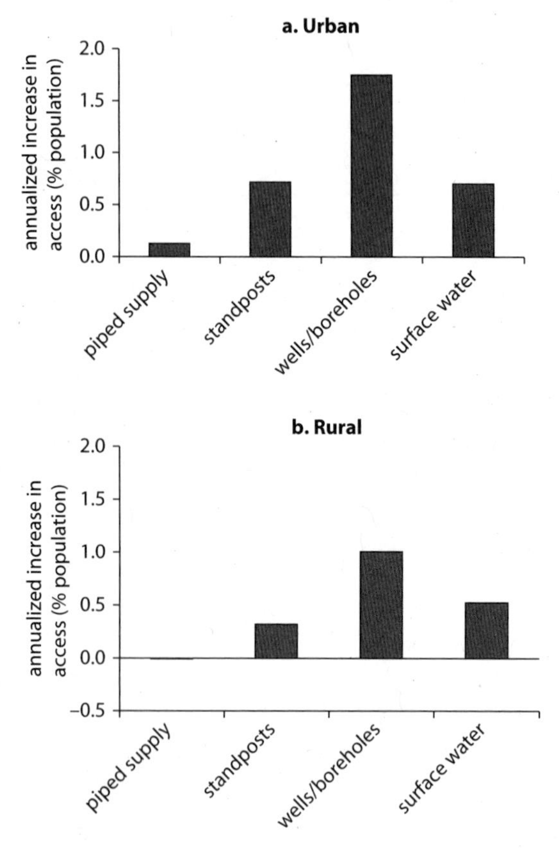

Source: Banerjee, Wodon, and others 2008.

Waterborne sewerage systems are rare in Africa. Only half of the large cities operate a sewerage network at all, and only in Namibia, South Africa, and the exceptional case of Senegal do some of the utilities covering the largest cities provide universal sewerage coverage. Little more than half of the households with piped water also have flush toilets, which are often connected to septic tanks rather than to sewers.

Patterns of access to sanitation vary dramatically across income groups. Open defecation is widely practiced in the lowest income quintile and not practiced at all in the highest. Conversely, improved latrines and septic tanks, virtually nonexistent among the poorest quintiles, are used by only 20–30 percent of the population in the

Figure 1.4 Coverage of Water Services, by Income Quintile

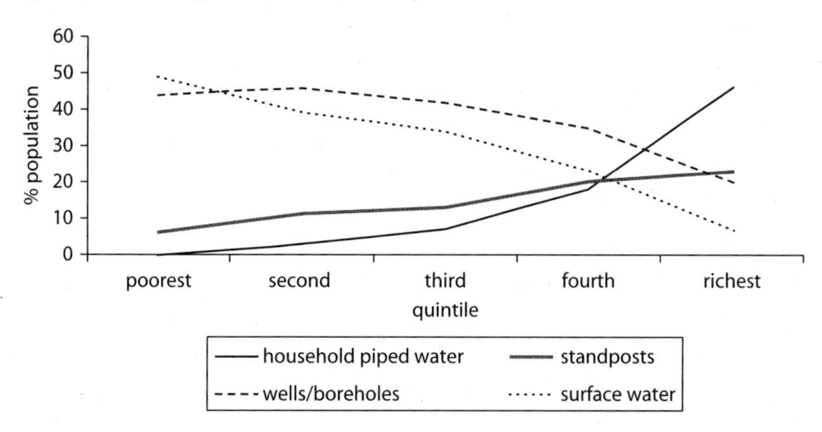

Source: Banerjee, Wodon, and others 2008.

Table 1.10 Patterns of Access to Sanitation in Africa
(percentage of population)

Area	*Open defecation*	*Traditional latrine*	*Improved latrine*	*Septic tank*
Urban	8	51	14	25
Rural	41	51	5	2
National	34	52	9	10

Source: Banerjee, Wodon, and others 2008.

richest. Access to improved latrines parallels that of septic tanks, suggesting that despite their lower cost, improved latrines remain something of a luxury, with little success in penetrating the middle of the income distribution. More important, the minimal presence of improved sanitation across poorer groups highlights a crucial issue—that high average rates of coverage do not help the most vulnerable populations. Traditional latrines are by far the most egalitarian form of sanitation, accounting across income ranges for about 50 percent of households (figure 1.5).

Traditional latrines are not only are the most common form of sanitation in Africa, but they are also the fastest growing. In recent years they have been used by an additional 2.8 percent of the population each year in urban areas and an additional 1.8 percent in rural areas, more

Figure 1.5 Coverage of Sanitation Services, by Income Quintile

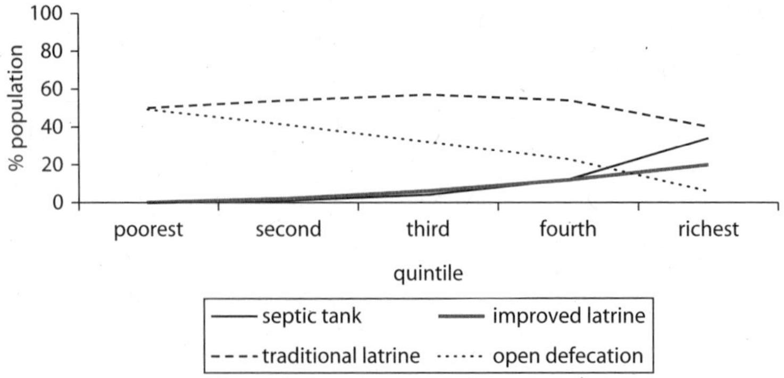

% population
100
80
60
40
20
0
poorest second third fourth richest
quintile
septic tank
improved latrine
traditional latrine
open defecation

Source: Morella, Foster, and Banerjee 2008.

than twice the rate of expansion of septic tanks and improved latrines combined (figure 1.6). Growth in the use of traditional latrines is concentrated among the poorer quintiles and of improved latrines and septic tanks among the richer quintiles. Because the MDG target focuses on the two most improved sanitation options, the expanding use of traditional latrines does not always fully register in policy discussions. Meanwhile, the prevalence of open defecation in Africa has finally begun to decline, albeit at a very modest pace.

Key Finding 3: High Costs, High Tariffs, and Regressive Subsidies

African water utilities operate in an environment of high costs, with two-thirds of the utilities operating in 2005 within the cost band of \$0.4 to \$0.8/m^3. Since then, costs have continued to rise in nominal terms. The high average cost of operations and maintenance (O&M) in Africa is somewhat misleading, driven as it is by the high cost of providing services in the middle-income countries of South Africa and Namibia, which is more than \$1, because it includes the cost of purchasing bulk water. Overall, Africa's experience in recovering operating costs is positive, with many utilities setting tariffs at levels high enough to recoup O&M costs. In fact, African tariffs are highest among the developing regions, with the operating ratio very close to 1 mainly because utilities spend everything they collect and nothing over that. Thus, they are not adequately funding either capital expenditures or rehabilitation or maintenance.

Figure 1.6 Annual Growth in the Use of Sanitation Types, 1990–2005
(percent)

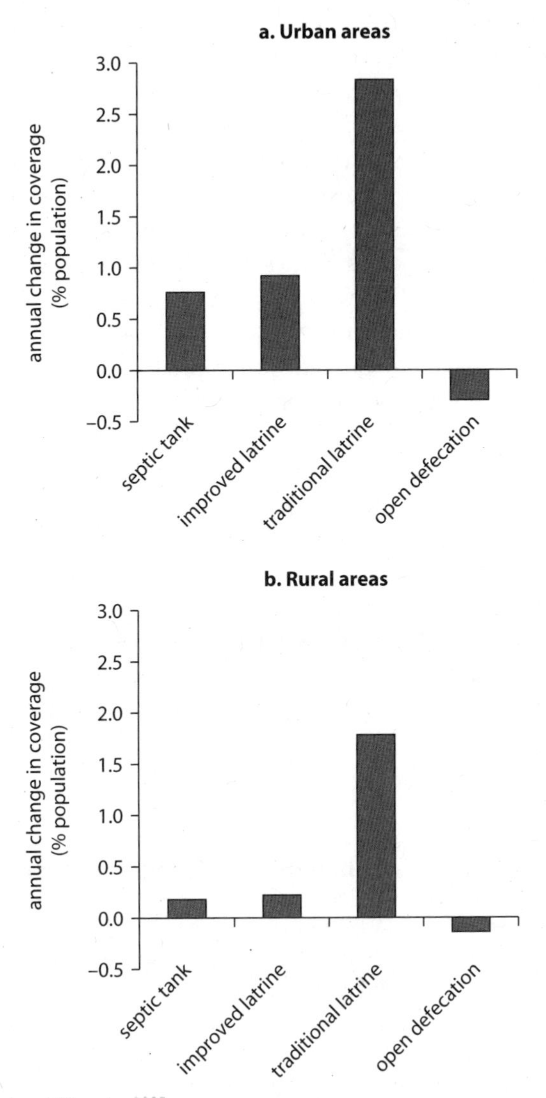

Source: Morella, Foster, and Banerjee 2008.

Full cost recovery is far off. Only four utilities in middle-income countries achieve their capital cost recovery at an average level of consumption of 10 m³/month. It is only in the last block of the increasing block tariff structure that prices are set with an eye to cost recovery because of the widespread perception that recouping capital costs from consumers is not feasible because of the limited budgets of African households.

In most countries of the region, utilities' capital costs have been almost entirely subsidized by the state or by donors, but the subsidies are highly regressive, especially those to residential consumers in urban areas. Across the bottom half of the income distribution, barely 10 percent of households have access to piped water. Indeed, more than 80 percent of households with piped water come from the top two quintiles of the income distribution. Because poorer households are almost entirely excluded, they cannot benefit from subsidies embedded in prices for piped water. In many cases, targeting performance is further exacerbated by poor tariff design, with widespread use of minimum charges and rising block tariffs that provide large blocks of highly subsidized water to all consumers.

Tariffs high enough to provide full capital cost recovery should be affordable for half of the population in Africa—and for about 40 percent of the population in low-income countries—but not for the remainder. Assuming household average consumption of 10 cubic meters a month (or about 65 liters per capita a day), a monthly utility bill under full-cost-recovery pricing of $1 would be about $10. Based on an affordability threshold of 5 percent of household income, full-cost-recovery tariffs would prove affordable for 40 percent of the population in low-income countries (figure 1.7). With about 10 percent of the national population already enjoying a direct water connection, an additional 30 percent of the population could be connected to water service and be able to pay for it. Most of the remaining 60 percent of the population would be able to afford bills of about $6 a month.

Key Finding 4: The Stark Challenge of Financing the MDG

The overall price tag for reaching the MDG target for access to WSS is estimated at $22.6 billion per year, or 3.5 percent of Africa's gross domestic product. Most of that sum is related to the water sector, which is estimated to require allocations up to $17 billion per year, or 2.7 percent of Africa's gross domestic product (GDP) (table 1.11). The cost of new infrastructure is the largest share, requiring allocations of up to 1.5 percent of Africa's GDP every year, or 43 percent of overall spending. O&M needs are the next largest category, standing at 1.1 percent of Africa's GDP, or 31 percent of overall costs. Rehabilitation of existing assets requires lower yet substantial allocations—up to 0.9 percent of Africa's GDP—accounting for one-fourth of the overall needs.

Figure 1.7 Affordability of Full-Cost-Recovery Tariffs in Low-Income Countries

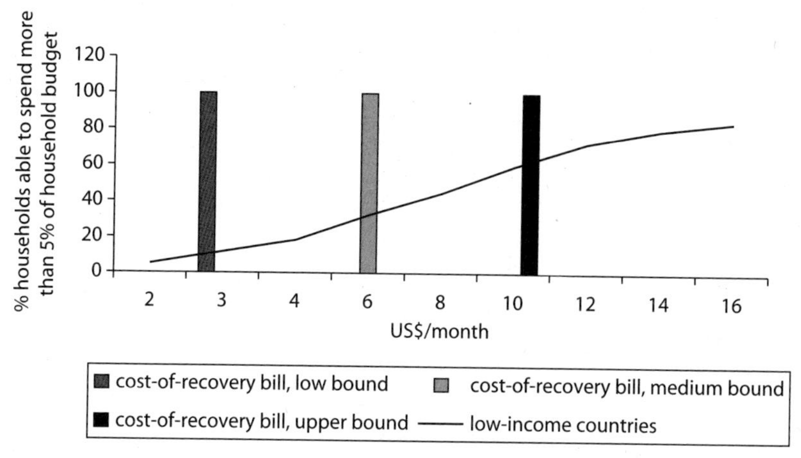

Source: Adapted from Banerjee, Wodon, and others 2008.

The composition of spending needs differs between middle- and low-income countries (table 1.12). Low-income countries (fragile or nonfragile) and resource-rich countries show much similarity, with costs divided almost equally among expansion, rehabilitation, and maintenance. Conversely, middle-income countries focus more on maintenance, which accounts for half of the overall spending needs, but the high coverage rates and relatively lower rehabilitation backlog make infrastructure expansion and rehabilitation less of a priority.

The affordability of meeting the MDG challenge appears to correlate strongly with a country's income. Halving the share of the population that lacks access to WSS services by 2015 is estimated to require only 1.5 percent of middle-income countries' GDP per year. Resource-rich countries would have to invest twice as much annually—3 percent of their GDP. The bill becomes prohibitively expensive for low-income countries, which would have to allocate at least 7 percent of their GDP to WSS every year to meet the goal. The burden would be even higher for fragile states: almost 12 percent of GDP each year.

As of 2005, Sub-Saharan Africa spends about $7.9 billion a year (1.2 percent of the region's GDP) on WSS—about a third of what is required if the MDG is to be met. In absolute terms, spending levels vary significantly across the country groups (table 1.13): Middle-income countries spend $2.6 billion, followed by low-income countries ($1.8 billion),

Table 1.11 Overall WSS Spending Needs

	Share of GDP (%)					*$ million per year*				
	CAPEX					*CAPEX*				
	Expansion	*Rehabilitation*	*Total CAPEX*	*O&M*	*Total needs*	*Expansion*	*Rehabilitation*	*Total CAPEX*	*O&M*	*Total needs*
Water	1.13	0.68	1.80	0.89	2.69	7,225	4,327	11,553	5,686	17,239
Sanitation	0.41	0.21	0.62	0.22	0.84	2,617	1,352	3,969	1,432	5,401
Total	1.54	0.89	2.42	1.11	3.53	9,843	5,679	15,522	7,118	22,640

Source: Authors' calculations.
Note: CAPEX = capital expenditure.

Table 1.12 Breakdown of Spending Needed to Meet MDGs in WSS, by Spending Category and Country Group

	Share of GDP (%)					*$ million per year*				
	CAPEX					*CAPEX*				
	New investment	*Rehabilitation*	*Total CAPEX*	*O&M*	*Total spending needs*	*New investment*	*Rehabilitation*	*Total CAPEX*	*O&M*	*Total spending needs*
Sub-Saharan Africa	1.5	0.9	2.4	1.1	3.5	9,843	5,679	15,522	7,118	22,640
Resource-rich countries	1.3	0.8	2.1	0.8	2.9	2,864	1,741	4,605	1,759	6,364
Middle-income countries	0.4	0.4	0.7	0.7	1.5	1,034	951	1,985	1,991	3,976
Low-income, fragile countries	5.9	2.7	8.5	3.3	11.8	2,208	1,006	3,213	1,223	4,437
Low-income, nonfragile countries	3.4	1.8	5.1	1.9	7.1	3,714	1,968	5,682	2,128	7,810

Source: Authors' calculations.
Note: CAPEX = capital expenditure.

Table 1.13 Spending by Functional Category, Annualized Average Flows, 2001–05

	Share of GDP (%)			*$ million per year*		
	O&M	*Total CAPEX*	*Total spending*	*O&M*	*Total CAPEX*	*Total spending*
Sub-Saharan Africa	0.5	0.7	1.2	3,112	4,778	7,890
Low-income, fragile countries	0.3	0.8	1.1	128	313	441
Low-income, nonfragile countries	0.3	1.4	1.7	307	1,533	1,840
Middle-income countries	0.7	0.2	1.0	1,996	641	2,637
Resource-rich countries	0.1	0.7	0.8	188	1,564	1,753

Sources: Briceño-Garmendia, Smits, and Foster 2008 for public spending; PPIAF 2008 for private flows; Foster and others 2008 for financiers from outside the Organisation for Economic Co-operation and Development.
Note: CAPEX = capital expenditure.

and resource-rich countries ($1.7 billion); fragile states spend about $0.5 billion in capital investment and O&M. Expressed as a percentage of GDP, infrastructure spending fluctuates widely across different country groups. Low-income countries and fragile states spend 1.1 and 1.7 percent of their GDP, respectively, whereas middle-income countries and resource-rich countries spend 1 percent or less of theirs (1.0 and 0.8 percent, respectively). The composition of spending also varies substantially across country groups. Middle-income countries allocate 80 percent of WSS spending to maintenance, reflecting the fact that they have already built much of the infrastructure needed. By contrast, the other country groups allocate no more than 30 percent to this item. Therefore, resource-rich countries, low-income countries, and fragile states spend 70 to 90 percent of their budgets for WSS infrastructure on capital investments. Although this reflects their need to build new facilities, a danger looms of neglecting the maintenance needs of the limited network that is available.

Inefficiencies of various kinds (incomplete execution of budgets, operational inefficiencies, and underpricing) total an estimated $2.9 billion a year (0.5 percent of GDP). Eliminating those efficiencies would provide a large share of the additional funds needed to achieve the MDG. Even if all the efficiency gains were realized, however, a funding gap would remain. Subtracting existing spending and potential efficiency gains from the spending needed to reach the MDG leaves an annual funding gap of about $11.9 billion a year, or 1.8 percent of GDP (table 1.14). Although the gap is widest for capital investment ($8.6 billion), a large shortfall also exists for O&M.

Table 1.14 Funding Gap
($ million per year)

	Total needs	Spending traced to needs	Gain from eliminating inefficiencies	Sources of inefficiency			(Funding gap) or surplus
				Underexecution of budget	Operating inefficiencies	Underpricing	
Sub-Saharan Africa	–22,640	7,890	2,877	168	1,259	1,450	–11,873
Low-income, fragile countries	–4,531	441	471	6	106	358	–3,620
Low-income, nonfragile countries	–7,810	1,840	685	39	265	381	–5,285
Middle-income countries	–3,987	2,637	1,037	8	492	537	–312
Resource-rich countries	–6,364	1,753	522	137	172	214	–4,089

Source: Briceño-Garmendia, Smits, and Foster 2008.

The smallest funding gap is found in middle-income countries, where inefficiencies are greatest. After tackling the inefficiencies, middle-income countries would face a negligible funding gap of $0.3 billion, most of which could be realized by reallocating resources from O&M to capital expenditure or from another infrastructure sector. The largest funding gap remains in low-income countries (nonfragile), which account for about half of the total funding gap for Sub-Saharan Africa ($5.3 billion).

In the aggregate, the region needs to increase capital investment in water infrastructure by 1.3 percent of GDP. Low-income, nonfragile countries need to invest an additional 3.3 percent and fragile states an additional 6.8 percent.

Key Finding 5: Institutional Reform for Better Water Sector Performance

Many African governments have reformed their WSS systems in the past two decades to provide better services for their citizens. Countries that have pursued institutional reforms have built more efficient and effective sector institutions and achieved faster expansion of higher quality services. The potential dividend of such efforts is large, because addressing utility inefficiencies alone could make a substantial contribution to closing the sector funding gap in many countries. Utilities that have decentralized their WSS services or adopted private sector management have done a better job of eliminating inefficiencies and other hidden costs than those that have not. Unbundling of services can also be beneficial, but unbundling is rare in Africa and exclusively concentrated in middle-income countries, whose superior performance can be explained for many other reasons. At the same time, higher levels of regulation and better governance of utilities (often accompanied by corporatization) are associated with lower efficiency (figure 1.8).

The reform agenda has had two major thrusts: increasing private participation and improving governance from within.

Private sector participation has helped to improve utility performance, with Senegal being particularly noteworthy. Management contracts awarded to private operating companies, being relatively short-term instruments, have had a material effect on improving revenue collection and service continuity, but they have not had much of an impact on more intractable issues, such as reducing unaccounted-for water and expanding access. Lease contracts have drastically improved access and boosted operational efficiency, but, except in Côte d'Ivoire, the associated investments

Figure 1.8 Hidden Costs and Institutions

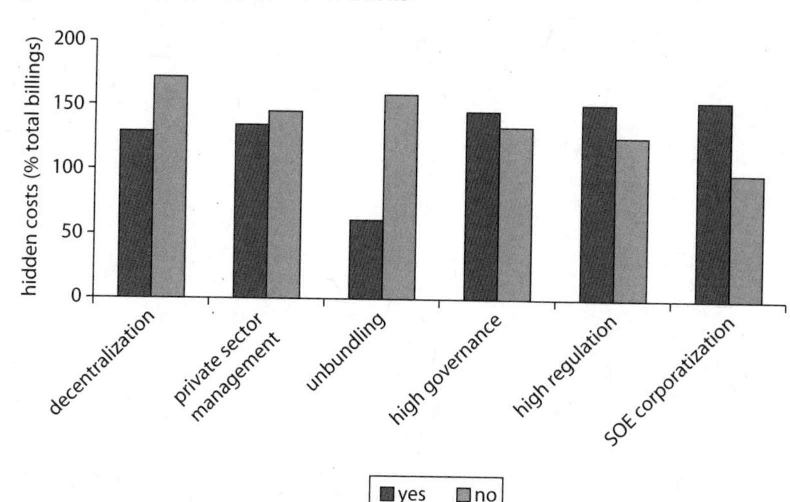

Source: Banerjee, Skilling, and others 2008.
Note: SOE = state-owned enterprise.

have been publicly financed. The lease contracts in Guinea and in Maputo have been affected by a lack of coordination between the private contractor and the government, which has stalled progress in some key areas, such as unaccounted-for water. Overall, private sector contracts accounted for almost 20 percent of the increase of household connections in the region, twice the amount that would be expected given their market share of only 9 percent (table 1.15). However, half of these gains were made in Côte d'Ivoire alone (which has been adversely affected since the onset of civil war in 2002).

About half of the countries (mainly anglophone) have established dedicated regulatory agencies for the water sector, although a significant number of these have not adopted private sector participation. Conversely, a number of francophone countries with private participation have adopted regulatory frameworks contractually, without establishing an independent regulatory agency. No evidence seems to support the superiority of any one of these two approaches. Even where explicit regulatory frameworks have been established, these typically meet only around half of the corresponding good practice criteria. However, evidence for the links between introducing an independent regulator and improving performance is negligible for the water sector. Similarly, no conclusive evidence is found of the superiority of regulation by contract over the traditional form of regulation by agency.

Table 1.15 Overview of Impact of Private Sector Participation on Utility Performance

Country or city	Contract	*Unit change in performance before and after private participation*					
		Household connections	*Improved water*	*Service continuity*	*Unaccounted-for water*	*Collection ratio*	*Labor productivity*
Gabon	Concession contract	+20			−8		
Mali		+15	+29		−14		
Côte d'Ivoire	Lease contract or *affermage*	+19	+22				+2.6
Guinea			+27		−0		
Maputo			+2	+10	−1	+24	
Niger		+9	+3		−5		+3.2
Senegal		+18	+17		−15		+2.8
Johannesburg	Management contract				−0	+10	
Kampala				+6	−2	+12	
Zambia				+5	−28	+19	

Source: Adapted from Marin 2009.
Note: Blank cells denote missing data; household connections and improved water are measured as additional percentage points of households with access; service continuity is measured as additional hours per day of service; unaccounted-for water is measured as lower percentages of lost water; collection ratio is measured as additional percentage points of collection; and labor productivity is measured as additional thousands of connections served per employee.

Of governance reforms that appear to be the most important drivers of higher performance, two are especially promising: performance contracts with incentives and independent external audits. For instance, Uganda's water company has had success using a performance contract that offers incentives for good performance and improves accountability. The introduction of independent audits has also positively affected efficiency.

A Multidimensional Snapshot of WSS in Africa

What policies are appropriate to deal with the state of the sectors just reviewed? How can WSS services be improved and access to them widened to include more of the continent's people? No recipe book neatly lays out the steps that each country should adopt to enhance coverage. In fact, the challenge of expanding access differs immensely across Sub-Saharan Africa, and so do the explanations for mixed performance.

The rest of this volume presents a snapshot of sector performance, financing resources, and institutional, regulatory, and governance frameworks that is meant to augment our understanding of specific country experiences, help define barriers and constraints, measure resources and capacities, and identify opportunities for improvement.

Chapters 2 and 3 set the stage by presenting access trends and market structures in water and sanitation sectors, respectively. Chapter 4 discusses the sector's organization and regulatory arrangements. An analysis of performance variables in urban water utilities follows in chapter 5. Tariff structures, subsidy mechanisms, and affordability themes are introduced in chapter 6. Chapters 7 and 8 present financing arrangements for WSS, estimate the amounts that will have to be spent to achieve the MDG targets for access to WSS, and calculate the gap between available financing and the amounts needed. Finally, chapter 9 provides menu of options that may be used to bridge the funding gap in water and sanitation. These concluding chapters also review policy options.

The chapters are supported by a comprehensive set of tabular appendixes that present the information base generated from AICD's extensive data-collection and data-processing efforts. Six sets of tables follow: Appendix 1 deals with access to WSS services (chapters 2–3). Appendix 2 relates to the institutional landscape (chapter 4). Appendix 3 is concerned with the technical and financial performance of water utilities (chapter 5). Appendix 4 relates to utility tariffs. Appendix 5 explores the affordability of WSS services (chapter 6). Appendix 6 deals with investment needs and the gap between those needs and available resources (chapters 7–8).

Annex 1.1 Surveys in the AICD DHS/MICS Survey Database

	Available observations			*Year of survey*		*Included in the trend analysis*
Country	*1990–95*	*1996–2000*	*2001–05*	*DHS*	*MICS*	
Benin		√	√	1996, 2001		X
Burkina Faso	√	√	√	1993, 1999, 2003		X
Cameroon	√	√	√	1991, 1998, 2004		X
Central African Republic	√			1995		
Chad		√	√	1997, 2004		X
Comoros		√		1996		
Congo, Dem. Rep.	√		√		2000	X
Congo, Rep.			√	2005		
Côte d'Ivoire	√	√		1994, 1999		X
Ethiopia		√	√	2000, 2005		X
Gabon		√		2000		
Ghana	√	√	√	1993, 1998, 2003		X
Guinea			√	1999, 2005		X
Kenya	√	√	√	1993, 1998, 2003		X
Lesotho		√	√	2005	2000	X
Madagascar	√	√	√	1992, 1997, 2004		X
Malawi	√	√	√	1992, 2000, 2004		X
Mali		√	√	1996, 2001		X
Mauritania			√	2001		
Mozambique		√	√	1997, 2003		X
Namibia	√	√		1992, 2000		X
Niger	√	√		1992, 1998		X
Nigeria	√	√	√	1990, 1999, 2003		X
Rwanda	√	√	√	1992, 2000, 2005		X
Senegal	√	√	√	1993, 1997, 2005		X
South Africa		√		1998		
Sudan		√			2000	
Tanzania	√	√	√	1992, 1999, 2004		X
Togo		√		1998		
Uganda	√		√	1995, 2001		X
Zambia	√	√	√	1992, 1996, 2002		X
Zimbabwe	√	√		1994, 1999		X

Source: Banerjee, Wodon, and others 2008.
Note: DHS = demographic and health survey, MICS = multiple-indicator cluster survey.

Annex 1.2 Surveys in the AICD Expenditure Survey Database

	Country	*Type and year of survey*	*Sample size*	*Questions on water supply*	*Questions on sanitation*
1	Angola	Integrated Expenditure Survey 2000	10,116	Yes	No
2	Benin	Core Welfare Indicators Questionnaire 2002	5,350	Yes	Yes
3	Burkina Faso	Core Welfare Indicators Questionnaire 2003	8,500	Yes	Yes
4	Burundi	Priority Survey 1998	6,668	Yes	No
5	Cameroon	Enquête Camerounaise auprès des ménages II 2001	4,584	Yes	Yes
6	Cape Verde	Integrated Expenditure Survey 2001	—	Yes	Yes
7	Chad	Enquête sur la consommation et le secteur informel au Tchad 2002	10,992	Yes	Yes
8	Congo, Dem. Rep.	Integrated Expenditure Survey 2005	10,801	Yes	Yes
9	Congo, Rep.	Enquête Congolaise auprès des ménages pour l'évaluation de la pauvreté 2005	12,097	Yes	Yes
10	Côte d'Ivoire	Integrated Expenditure Survey 2002	5,002	Yes	Yes
11	Ethiopia	Welfare Monitoring Survey 2000	16,672	Yes	Yes
12	Gabon	Core Welfare Indicators Questionnaire 2005	7,902	Yes	Yes
13	Ghana	Ghana Living Standards Survey 1998/99	5,991	Yes	Yes
14	Guinea-Bissau	Core Welfare Indicators Questionnaire 2002	3,216	Yes	Yes
15	Kenya	Welfare Monitoring Survey 1997	10,874	Yes	Yes
16	Madagascar	Enquête prioritaire des ménages 2001	5,081	Yes	Yes
17	Malawi	Integrated Household Survey 2003	11,280	Yes	Yes
18	Mauritania	Enquête permanente sur les conditions de vie des ménages 2000	5,865	Yes	Yes
19	Morocco	Integrated Household Survey 2003	5,129	Yes	Yes
20	Mozambique	Inquérito aos agregados familiares sobre orçamento familiar 2002/03	8,703	Yes	Yes
21	Niger	Integrated Household Survey 2005	6,690	Yes	Yes
22	Nigeria	Nigeria Living Standards Survey 2003	19,158	Yes	Yes
23	Rwanda	Enquête intégrale sur les conditions de vie des ménages (avec module budget et consommation) 1999	6,420	Yes	Yes
24	São Tomé and Príncipe	Enquête sur les conditions de vie des ménages 2000	6,594	Yes	Yes

(continued next page)

	Country	Type and year of survey	Sample size	Questions on water supply	Questions on sanitation
25	Senegal	Integrated Expenditure Survey 2001	2,418	Yes	Yes
26	Sierra Leone	Integrated Household Survey 2003	3,713	Yes	Yes
27	South Africa	Integrated Expenditure Survey 2000	26,263	Yes	Yes
28	Tanzania	Household Budget Survey 2000	22,207	Yes	Yes
29	Uganda	National Household Survey 2002	9,710	Yes	Yes
30	Zambia	Living Conditions Monitoring Survey 2002	9,715	Yes	Yes
Total			267,711	30	28

Source: Banerjee, Wodon, and others 2008.
Note: — = not available.

Annex 1.3 Introducing a Country Typology

Africa's numerous countries face widely diverse economic situations. Understanding that structural differences in countries' economies and institutions affect their growth and financing challenges as well as their economic decisions (Ndulu and others 2007), this chapter introduces a four-way typology to organize the rest of the discussion. This typology provides a succinct way of illustrating the diversity of infrastructure financing challenges faced by different African countries.

Middle-income countries have a gross domestic product (GDP) per capita in excess of $745 but less than $9,206. Examples include Cape Verde, Lesotho, and South Africa (World Bank 2007).

Resource-rich countries are countries whose behaviors are strongly affected by their endowment of natural resources (IMF 2007).[2] Resource-rich countries typically depend on minerals, petroleum, or both. A country is classified as resource rich if primary commodity rents exceed 10 percent of GDP. (South Africa is not classified as resource intensive, using this criterion.) Examples include Cameroon, Nigeria, and Zambia.

Fragile states are low-income countries that face particularly severe development challenges, such as weak governance, limited administrative capacity, violence, or the legacy of conflict. In defining policies and approaches toward fragile states, different organizations have used differing criteria and terms. Countries that score less than 3.2 on the World Bank's Country Policy and Institutional Performance Assessment belong to this group. Fourteen countries in Africa are in this category. Examples include the Democratic Republic of Congo, Côte d'Ivoire, and Sudan (World Bank 2005).

Other low-income countries compose a residual category of countries with GDP per capita below $745 and that are neither resource-rich nor fragile states. Examples include Benin, Ethiopia, Senegal, and Uganda.

Notes

1. See United Nations, "Millenium Development Goals, http://www.un.org/millenniumgoals/.
2. See also Paul Collier and Stephen O'Connell, draft chapter (2006) for the synthesis volume of the African Economic Research Consortium's *Explaining African Economic Growth* project, Oxford University and Centre for Study of African Economies, and Swarthmore College and Centre for Study of African Economies.

Bibliography

Banerjee, S., H. Skilling, V. Foster, C. Briceño-Garmendia, E. Morella, and T. Chfadi. 2008. "Ebbing Water, Surging Deficits: Urban Water Supply in Sub-Saharan Africa." AICD Background Paper 12. World Bank, Washington, DC.

Banerjee, S., Q. Wodon, A. Diallo, N. Pushak, H. Uddin, C. Tsimpo, and V. Foster. 2008. "Access, Affordability and Alternatives: Modern Infrastructure Services in Sub-Saharan Africa." AICD Background Paper 2. World Bank, Washington, DC.

Briceño-Garmendia, C., K. Smits, and V. Foster. 2008. "Financing Public Infrastructure in Sub-Saharan Africa: Patterns and Emerging Issues." AICD Background Paper 15. World Bank, Washington, DC.

Esrey, S. A., J. B. Potash, L. Roberts, and C. Shiff. 1991. "Effects of Improved Water Supply and Sanitation on Ascariasis, Diarrhea, Dracunculiasis, Hookworm Infection, Schistosomiasis and Trachoma." *Bulletin of the World Health Organization* 69 (5): 609–21.

Foster, Vivien, and Cecilia Briceño-Garmendia, eds. 2009. *Africa's Infrastructure: A Time for Transformation*. Paris and Washington, DC: Agence Française de Développement and World Bank.

Foster, Vivien, William Butterfield, Chuan Chen, and Nataliya Pushak. 2008. "Building Bridges: China's Growing Role as Infrastructure Financier for Sub-Saharan Africa." Trends and Policy Options 5, Public-Private Infrastructure Advisory Facility, World Bank, Washington, DC.

Hutton, G., and L. Haller. 2004. "Evaluation of the Costs and Benefits of Water and Sanitation Improvements at the Global Level." World Health Organization, Geneva.

IMF (International Monetary Fund). 2007. "Regional Economic Outlook: Sub-Saharan Africa." International Monetary Fund, Washington, DC.

JMP (Joint Monitoring Programme). 2000. *Global Water Supply and Sanitation Assessment 2000 Report*. Geneva: World Health Organization; New York: United Nations Children's Fund.

———. 2008. *Progress on Drinking Water and Sanitation. Special Focus on Sanitation*. Geneva: World Health Organization; New York: United Nations Children's Fund.

Keener, S., M. Luengo, and S. G. Banerjee. 2009. "Provision of Water to the Poor in Africa: Experience with Water Standposts and the Informal Water Sector." AICD Working Paper 13. World Bank, Washington, DC.

Marin, P. 2009. "Public Private Partnerships for Urban Water Utilities: A Review of Experiences in Developing Countries." Trends and Policy Options 8, Public-Private Infrastructure Advisory Facility and World Bank, Washington, DC.

Morella, E., V. Foster, and S. Banerjee. 2008. "Climbing the Ladder: The State of Sanitation in Sub-Saharan Africa." AICD Background Paper 13. World Bank, Washington, DC.

Ndulu, Benno J., Stephen A. O'Connell, Robert H. Bates, Paul Collier, and Charles C. Soludo, eds. 2007. *The Political Economy of Economic Growth in Africa, 1960–2000*. Volume 1. Cambridge: Cambridge University Press.

PPIAF (Public-Private Infrastructure Advisory Facility). 2008. "Private Participation in Infrastructure Project Database." http://ppi.worldbank.org/.

World Bank. 2005. "Infrastructure Finance for Africa—A Strategic Framework." Unpublished concept note, World Bank, Washington DC.

———. 2007. "DEPweb Glossary." Development Education Program, World Bank, Washington, DC. http://www.worldbank.org/depweb/english/modules/glossary.html#middle-income.

CHAPTER 2

Access to Safe Water: The Millennium Challenge

The water landscape in Africa is characterized by discrepancies within and among countries. Some countries are closer than others to achieving the water target spelled out in the Millennium Development Goals (MDGs).[1] In this chapter, we present the recent evolution and current status of water service in Africa, focusing on the underlying markets—urban formal, urban informal, and rural—each with their unique attributes and players. Some countries emerge as robust performers in expanding coverage in urban and rural areas, whereas others have remained stagnant or fallen behind in serving their population.

The Importance of Wells and Boreholes in Water Supply

Less than one-third of African households have reached the top parts of the ladder. About 15 percent of African households receive piped w[illegible] through household connections; another 15 percent receive i[illegible] standposts. Wells and boreholes cover 37 percent of ho[illegible] them the most prevalent form of water supply in t[illegible] remainder of the population relies on surface wat[illegible] urban areas, water vendors serve about 2 percent o[illegible]

Rates of water supply coverage show tremendous heterogeneity from one country to another. The variation in household piped-water coverage is wide—from 2 percent in Uganda to about 60 percent in South Africa. In most countries, piped water reaches less than 20 percent of households (figure 2.1a). Only three countries—Gabon, Senegal, and South Africa—can claim a piped-water coverage rate of more than 40 percent. The coverage of wells/boreholes and surface water reveal even greater variation (figure 2.1b).

The low rate of piped-water coverage reflects Africa's relatively low rate of urbanization. Piped-water coverage in rural areas is several magnitudes lower than in urban areas. Only 4 percent of rural households in Africa receive piped water, compared with 38 percent in urban areas. When public standposts are included, more than 60 percent of Africa's urban households have access to some kind of utility provided water. In rural areas, wells and boreholes and surface water predominate. More than 80 percent of Africa's rural households receive their water from these sources.

Richer households are much more likely to enjoy access to piped water than are poorer households. On the water-supply ladder, rising income is associated with piped water and a declining dependence on wells, boreholes, and surface water. In the lowest three quintiles of the wealth distribution, access to piped water through a household connection is well below 10 percent, with negligible coverage of the poorest households (table 2.1). Even in the fourth quintile, access to piped water within the household is less than 20 percent, whereas for the richest quintile it is close to 50 percent—still far from universal (and highly variable across countries).

Most of the countries in the sample of the Africa Infrastructure Country Diagnostic (AICD) are low-income countries with a per capita gross domestic product (GDP) of less than $1,000 per year, but the sample also includes several middle-income countries: Cape Verde, Gabon, Lesotho, Namibia, and South Africa. The degree of urbanization varies widely in Africa—from 12 percent in Uganda to 80 percent in Gabon—the average being about 35 percent.

Both income and urbanization are directly correlated with access to safe water. Higher incomes make safe water more affordable, and the greater population densities associated with urbanization help to reduce he cost of expanding access to modern services.

Access to piped water is four times greater in middle- than in low- e countries and three times greater in the most urbanized

Figure 2.1 African Households' Access to Various Forms of Water Supply

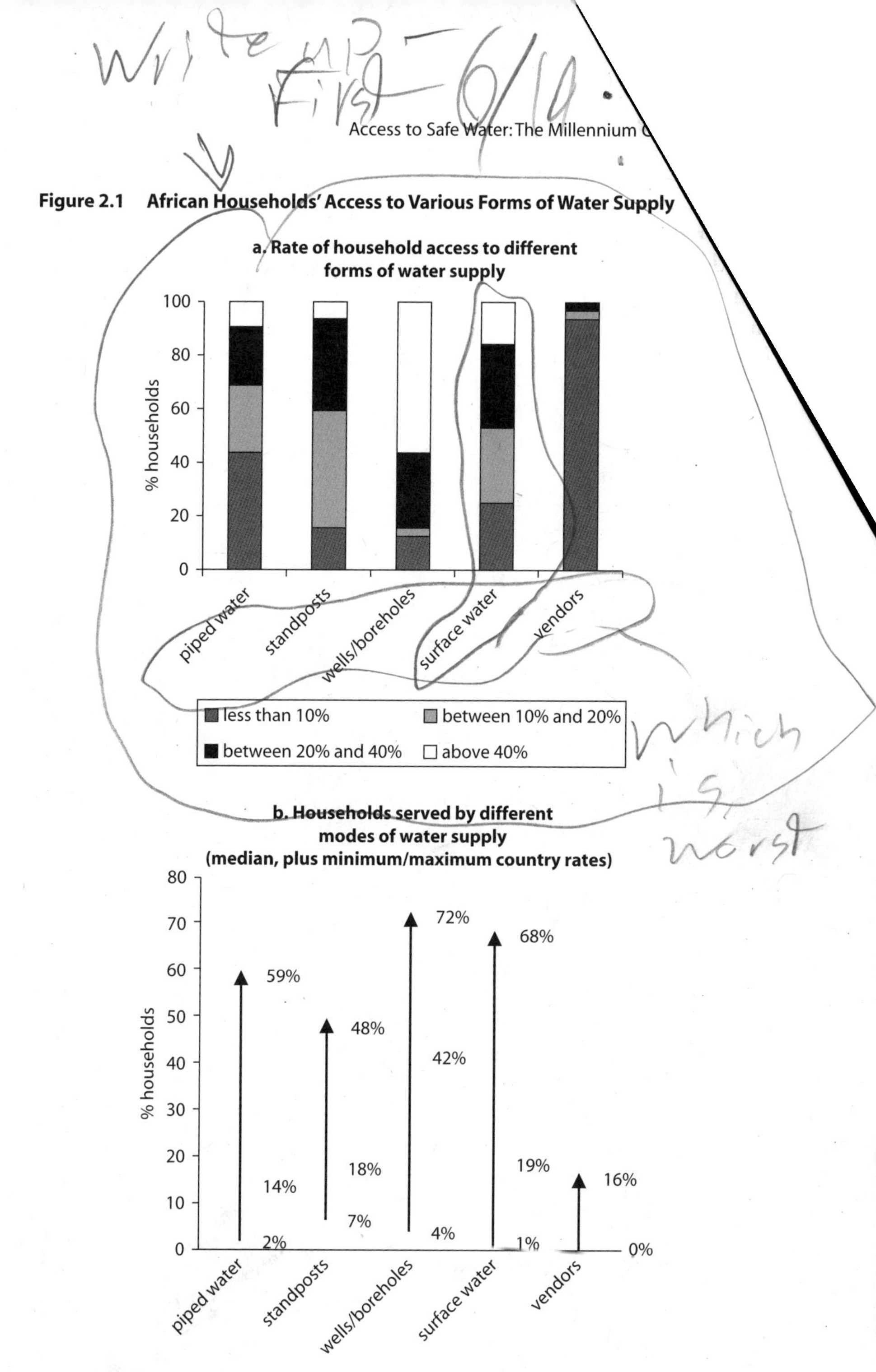

Source: Banerjee, Wodon, and others 2008.

Supply

				Quintile 1	Quintile 2	Quintile 3	Quintile 4	Quintile 5
			38	0	3	7	18	46
		10	25	6	11	13	20	23
	37	43	24	44	46	42	35	20
water	30	41	7	49	39	34	23	7
endors	2	1	4	1	1	2	2	2

Source: Banerjee, Wodon, and others 2008.

economies than in the least, and recourse to surface water is about twice as prevalent in the low-income and least urbanized countries than in the middle-income and most urbanized countries (table 2.2). These patterns hold across urban and rural service segments, and across the different quintiles of the distribution of spending on water service as well. Thus, in more highly urbanized countries, even the rural population is substantially better off. Nevertheless, even in middle-income and urbanized countries, the benefits of access are largely confined to the top three quintiles of the distribution, with too many in the bottom two quintiles still without access to safe water.

In the vast majority of countries the distribution of access is even more unequal than the distribution of income, exacerbating inequalities in society as a whole. Furthermore, the distribution of new connections resulting from the service expansions that have occurred in recent years is also more unequal than income. It appears, therefore, that the benefits of current access and new extensions tend to accrue to the better-off. This may be because access rates in Africa remain low even among the wealthier segments of the population, so it makes business sense for the utilities to initially concentrate their expansion efforts (Diallo and Wodon 2005).

Even if one controls for income and urbanization, some countries stand out as having much higher (or lower) levels of coverage than might be expected, and these cases merit closer study (figure 2.2). As seen in figure 2.2, Cameroon and Ghana have relatively high incomes and high rates of urbanization, yet their piped-water coverage is relatively small, suggesting underperformance. Senegal, by contrast, has coverage that compares favorably with that of peers at similar (and even greater) levels of income and urbanization. Sitting just to the left of Senegal, on the 50 percent urbanization line, Nigeria stands out as having low levels of